Marcelo Azevedo de souza
Alline de Almeida

Maths in Construction

Marcelo Azevedo de souza
Alline de Almeida

Maths in Construction

The best seller

ScienciaScripts

Imprint

Any brand names and product names mentioned in this book are subject to trademark, brand or patent protection and are trademarks or registered trademarks of their respective holders. The use of brand names, product names, common names, trade names, product descriptions etc. even without a particular marking in this work is in no way to be construed to mean that such names may be regarded as unrestricted in respect of trademark and brand protection legislation and could thus be used by anyone.

Cover image: www.ingimage.com

This book is a translation from the original published under ISBN 978-613-9-73062-9.

Publisher:
Sciencia Scripts
is a trademark of
Dodo Books Indian Ocean Ltd. and OmniScriptum S.R.L publishing group

120 High Road, East Finchley, London, N2 9ED, United Kingdom
Str. Armeneasca 28/1, office 1, Chisinau MD-2012, Republic of Moldova, Europe
Printed at: see last page
ISBN: 978-620-7-89638-7

Alline de Almeida

Graduated in Pedagogy from the University of Vale do Itajaí (UNIVALI), in the state of Santa Catarina, in the city of Itajaí. Specialised in School Management at AUPEX, in the state of Santa Catarina, in the city of Itajaí.

Marcelo Azevedo de Souza

He graduated in Mathematics from the Technical and Higher Education Centre of Western Paraná (CTESOP), in the state of Paraná, in the city of Assis Chateaubriand. He specialised in School Administration at the Universitas Lucius Annaeus Seneca (Unilas) in Itapema and holds a Master's degree in Natural Sciences and Mathematics from the University of Blumenau.

This work is the result of Marcelo Azevedo de Souza's dissertation in natural sciences and maths at the University of Blumenau (FURB) - Blumenau/Santa Catarina - Brazil.

The main purpose of this book is to show readers, teachers and students of youth and adult education and ethnomathematics researchers that the everyday mathematics used in construction can be related to the mathematics taught in schools and that the two can move forward in an organised way, with the aim of developing excellence in the teaching and learning of this subject.

The subjects chosen in this book are varied and based on students' everyday lives, so they can be applied in any classroom in primary and secondary education.

The book presents a brief introduction, explaining the choice of topic, general and specific objectives, the structure of each chapter, followed by creative and challenging activities, a timetable for application, research methods, data collection, resources to be used, mathematical construction games, and finally, tips for future application.

SUMMARY

INTRODUÇÃO

Throughout our careers as maths teachers, we have developed activities that we consider relevant and that go beyond the content required by the teaching programmes. We try to develop teaching that is meaningful to the students and that leads them to question and problematise the content and to a deeper understanding of the world of which they are a part. In this sense, I was already concerned about how to relate school mathematics content to different cultures, but in order to do this, I needed to propose the practice of teaching mathematics in a contextualised way.

Through informal conversations with students in Youth and Adult Education (EJA), some of them construction workers, we diagnosed difficulties in relating school maths to their professional practices. To change this scenario, we sought out challenging activities, always associating everyday contexts with pedagogical practices that would arouse the students' curiosity.

So the search for new methodological procedures continued, as I was looking for those that were meaningful, with the aim of establishing meanings for the teaching process in my classes.

In this sense, as a teacher-researcher, we are seeking, through this book, to socialise and deepen questions about the teaching of mathematics for Youth and Adult Education and, specifically, about concepts related to Ethnomathematics.

Youth and Adult Education (EJA) is a long-standing concern in Brazilian education. Most EJA students are people who are in the labour market and need a teaching process that takes into account their life experiences.

In his research on the EJA, Casério (2003, p. 60) argues that:

> [...] the adult learner is a worker, not a child, they are in the labour market, or aspire to it, and have to prepare themselves to enter it, so the content to be taught at school should refer both to the adult worker's life experience and to the formal content that explained this reality reflected by them.

Thus, the search for situations that take into account and value the students' everyday experiences and knowledge becomes fundamental to teaching and learning in the EJA. Through these everyday situations, it is possible to establish a relationship between school life and the students' working practices. According to Corrêa (2000), in the world of work its meanings are present, imbued with objectivity and subjectivity, whose significance is observed in the competence that the person shows when dealing with certain situations. The teaching of maths allows us to develop the skills and abilities that students need in their everyday lives, especially in the EJA, such as budgeting in a construction site, selling, exchanging and buying various goods. In this sense, school maths takes on its true meaning in the social and economic life of the EJA student.

Article 37 of Law No. 9.394/1996, which establishes the Guidelines and Bases of National Education (LDB), states that education for young people and adults is aimed at those who have not had access to or continued their studies. When they return to school, many EJA students feel fearful about school content, especially maths. They are faced with "teaching based on memorising rules or strategies for solving problems, or centred on content that is of little significance to the students, which certainly does not contribute to a good mathematical education" (BRASIL, 2002, p.5). As a result of teaching based on memorisation and repetition, EJA students end up giving up on school life for various reasons, including a lack of understanding of the content being taught or fear of facing the difficulties imposed by the school.

In order for EJA teaching to make sense in students' lives and establish a link between daily life and school life, we need to modify pedagogical practices by making sense of the content covered in class. EJA students generally bring with them professional and (cultural and historical) mathematical experiences, and are constantly involved in mathematical activities that, even if they don't recognise them in this way, involve aspects of common sense. It is therefore essential that teachers look for diversified strategies that allow students to participate in the construction of their own knowledge, thus promoting the creation of innovative teaching strategies and ideas linked to everyday practices (NICOLODI, 2011). As such, teachers can make use of these experiences to challenge and stimulate students. According to Freire (2006, p. 80):

> [...] the more students are problematised as beings in the world and with the world, the more they will feel challenged. The more they are challenged, the more they are obliged to respond to the challenge. Challenged, they understand the challenge in the very action of grasping it. But precisely because they grasp the challenge as a problem in its connections with others, on a level of totality and not as something petrified, the resulting understanding tends to become increasingly critical, and therefore increasingly disalienated.

Changes are needed in the regular teaching of maths, diversified activities that establish connections between school mathematical knowledge and students' experiences. Thus, we believe that some changes could be significant for the EJA, such as the application of activities related to the students' work practice.

As we work in the EJA with professionals who have experience in the field of construction, we seek to make the teaching of mathematics more flexible and contextualised by valuing and using the mathematics used in their professional activity to teach school mathematics, guided by Ethnomathematics. For Ubiratan D'Ambrosio (1990, p. 5) Ethnomathematics is:

> the art or technique of explaining, knowing and understanding in various cultural contexts. [...] it is a programme that aims to explain the processes of generating, organising and transmitting knowledge in diverse cultural systems and the interactive forces that act in and between the three processes [...]

Ethnomathematics seeks to value the mathematical, intuitive and cultural knowledge of the groups we work with. Therefore, we will seek to understand how mathematics is carried out in the working practices of construction workers, using their cultural base as a starting point for the insertion of school mathematics content. This is the development of a methodological proposal that encourages the construction of knowledge by students based on their experiences. Furthermore, by bringing school knowledge closer to the context, we can help overcome the prejudice that maths is an area of knowledge that is difficult to understand.

The mathematics experienced by construction workers is made up of observation and repetition of everyday activities, but they don't realise the importance of school mathematics in their daily practices. In this sense, the maths taught in the EJA establishes "connections" and "meanings" between school maths and *the* students' reality.

From this context, this research was developed, the central question of which is: What are the contributions of Ethnomathematics to a maths teaching process for young people and adults that links everyday building practices and the teaching of school maths?

Based on this research question, we defined the general objective of this work: To identify the contributions of Ethnomathematics to a maths teaching process for young people and adults, which linked everyday construction practices and the teaching of school maths.

To answer the general objective, we have set the following specific objectives:

- To identify mathematical concepts used by the students surveyed in the construction industry.
- Using mathematical concepts used in construction to learn school maths concepts.
- Describe and analyse activities carried out on the mathematical operations used in construction.
- Identify the principles of Ethnomathematics present in the pedagogical practice applied to the teaching of school maths.

In order to expand our knowledge of Ethnomathematics, we first sought to find out about published works on the subject. We consider the works presented below to be essential sources for the development of the research, because it is through this action that we gather important information. To do this, we conducted a survey of different databases that report scientific publications in the area of mathematics teaching, electronic journals on mathematics and ethnomathematics, the CAPES portal, the websites of postgraduate programmes at public and private universities such as: Universidade Regional de Blumenau (FURB), Universidade Federal de Santa Catarina (UFSC), Universidade Estadual de Campinas (UNICAMP), Unidade Integrada Vale do Taquari de Ensino Superior (UNIVATES) and Universidade Estadual Paulista (UNESP), Universidade do Vale do Rio dos Sinos (UNISINOS) and Universidade Aberta - Portugal (UA). We used the following terms as keywords: Ethnomathematics, Maths Education, Civil Construction and

Youth and Adult Education, and which are related to the object of study in this research, having been published in the period from 2000 to 2015.

Through research into books, dissertations and publications of papers and articles in scientific journals, it was possible to investigate the theoretical and methodological references of research into Ethnomathematics. Careful reading of various publications provided a reference for decisions on the theoretical-methodological definition and served as support during the preparation and systematisation of this book.

This book is organised into six chapters: Chapter I presents the introduction (choice of theme, general and specific objectives). Chapter II presents Maths Teaching and Youth and Adult Education in Brazil: a Historical Retrospective and EJA: Teaching and Learning.

Chapter III discusses ethnomathematics and its contributions to the teaching of school maths. In Chapter IV we outline the research procedures, characterise the participants, the research setting, the data collection instruments, the analysis of the results, as well as the mathematical concepts used in construction by EJA students, the ethnomathematics activities and games developed, the ethnomathematics principles present in teaching practice and the relationship between school mathematics and everyday mathematics. In addition, in the Final Considerations we present the possible contributions of Ethnomathematics to the teaching of school maths to young people and adults with professional experience in the construction industry. Completing the structure of this book are the references.

2

Looking back over the history of mankind, there have been various concerns in the daily lives of human beings, including: the search for food for survival, the search for shelter, the fight to defend oneself from predators and the search for explanations for the inexplicable, such as phenomena of nature and death. According to Ubiratan D'Ambrosio, what characterises the human being is the permanent search for "survival" and "transcendence":

> In the human species, the question of survival is accompanied by that of transcendence: the "here and now" is extended to the "where and when". The human species transcends space and time beyond the immediate and the sensible. The present extends into the past and the future, and the sensible expands into the remote. Human beings act according to their sensory capacity, which responds to the material [artefacts], and their imagination, often called creativity, which responds to the abstract [mindfacts] (D'AMBROSIO, 2004, p. 28).

Thus, understanding the history of humanity also means understanding how mathematical knowledge was applied at different times. However, several past civilisations developed specific mathematical knowledge, such as the Egyptians, Greeks and especially the European continent. The Europeans played a major role in maths, and even today you can find their contributions in our mathematical teaching system, such as Arithmetic and Euclidean Geometry. In Brazil, during the 1940s and 1950s, their influences were important for the development of maths, as several mathematicians who lived in Europe migrated to Brazil to teach maths (SILVA, 2007). The first Brazilian reports on the teaching of maths date back to 1810, when the Royal Military Academy was set up in Rio de Janeiro, which became the Polytechnic School in 1974.

However, the way we think has changed. Based on these changes, it is necessary to develop maths teaching in the direction of diversified and multidisciplinary teaching methodologies, with content linked to students' everyday lives. In this way, maths teaching would no longer be a process of memorising formulas and concepts, but would "identify the type of information that is appropriate for a given situation and conditions" (D'AMBROSIO, 1986, p. 14). In this way, it would be possible to study mathematical content or methods involving the students' reality, thus enabling a better quality of life for human beings.

But why study or teach maths? D'Ambrosio (1986, p. 44) states that:

> the true spirit of mathematics is the ability to model real situations, codify them appropriately, in such a way as to allow the use of known techniques and results in another, new context. In other words, the transfer of learning resulting from a certain situation to a new situation is a crucial point in what could be called learning maths, and perhaps the greatest goal of its teaching

The conclusion is that maths teaching can be applied in a variety of social, economic and cultural contexts. However, defining a teaching strategy is crucial to the quality of learning, and at this point, the role of the teacher is essential, as it is they who guide the student in mathematical learning (D'AMBORSIO, 1986). Therefore, the role of teachers is fundamental in motivating and adapting a methodology that relates students' daily experiences to the curriculum objectives.

Currently, in our school systems, mathematics still suffers from a lack of motivation and contextualisation in teaching. To change this scenario, D'Ambrosio (1986) showed that maths should be treated in a historical and dynamic way, relating current problems to students' interests. However, for this to be discussed in a broader way, it is necessary to apply a research rationale to each object of study. In general, maths in schools is not always worked on according to some of its purposes. Students end up only being concerned with learning abstract knowledge. As a result, they forget an important aspect such as: knowing what contributions maths will have in their lives and how maths should be taught and understood throughout students' lives, thus enabling a creative way of thinking about and reconstructing maths in our country.

In order for maths to be discussed in any environment in a meaningful way, modern movements emerged in Brazil that envisaged an improvement in the teaching of maths. Among them were the Brazilian Society for Mathematics Education (SBEM), the Group for Studies and Research in Mathematics Education (GEPEM), the National Meeting for Mathematics Education (ENEM) and the first International Seminar for Research in Mathematics Education (SIPEM), as well as some postgraduate programmes. Through these movements, maths education began to be innovated, aiming for quality in Brazilian teaching, thus generating creative and innovative ideas.

These movements have provided new opportunities throughout Brazil for thinking and acting in education systems, which still had a decentralised curriculum and focused on solving and memorising activities. Thus, innovative ideas with new methodological practices are due to good teacher training in basic and higher education, thus enabling classrooms to become a learning environment, related to their daily lives. Maths education makes this interaction possible because it has a transdisciplinary and innovative character (D'AMBROSIO, 1986

As such, maths education is a diverse and broad field for any study. We need maths to carry out daily operations, such as paying bills, transferring money, crossing the street, adding, subtracting, dividing and others. That's why it's necessary for academic maths to relate to the maths that each student already has during their life history. In this way, teachers and students will enjoy maths with meaning and significance for teaching and learning maths.

2.1 YOUTH AND ADULT EDUCATION IN BRAZIL: A HISTORICAL RETROSPECTIVE

Youth and adult education in our country originated around 1876 in Empire Brazil. During this period, "a number of educational reforms began to take place and these advocated the need for evening education for illiterate adults" (PORCARO, 2007, p. 1). Some of these reforms led Minister José Bento da Cunha Figueiredo, in the same year, to produce a report revealing that there were 200,000 students attending evening classes.

In 1930, basic adult education began to take place in schools to meet a possible demand that began to emerge as a result of industrialisation and consolidated a public elementary education system. According to Ribeiro (1997), this period was important for Brazilian society, which was undergoing major changes due to the concentration of population in urban centres and the process of industrialisation. Thus, the development of adult education at the beginning of the 20th century was slow but growing, mainly to meet the needs of professionals who were seeking access to professional training in order to enter the labour market.

Thus, industrialisation brought different points of view in relation to adult education, such as: valuing the mastery of spoken and written language, with a view to mastering production techniques; the acquisition of reading and writing as a tool for social advancement and adult literacy seen as a means of progress for the country (PORCARO, 2007). However, there was still a need for other reforms in EJA education, including guaranteeing free access to public schools for young people and adults.

But it wasn't until 1934 that the Constitution guaranteed the right and access to education for all. Free education then began to be offered in Brazil, reaching different contexts: social, economic, political and cultural. Consequently, the Federal Government, in partnership with the municipalities, established rules and competences to be fulfilled, such as "free comprehensive primary education and compulsory attendance for adults" (BRASIL, 1934, Art. 150, sole paragraph, point a). In this way, EJA students went back to school.

In 1947, the government launched the 1ª Adult Education Campaign, which aimed to provide three months of literacy and the other two periods of seven months focused on primary education, professional training and personal development of students, where "the campaign achieved significant results, articulating and expanding existing services and extending them to the various regions of the country" (RIBEIRO, 1997, p. 20). In this way, it was possible to allocate a period of time to EJA teaching and also to provide adequate and standardised training throughout Brazil.

With the success of the implementation, in 1947 the government created "specific teaching material for reading and writing for adults, which was distributed on a large scale by all the supplementary schools in the country" (NICOLODI, 2011, p. 23), through this distribution of materials it was possible to establish a teaching standard throughout Brazil.

However, discussions about illiteracy and adult education were expanded in Brazil. At the time, illiteracy was seen as something that hindered Brazilian development (CUNHA, 1999). In addition, illiterate adults were identified as psychologically and socially incapable and marginalised, subjected to economic, political and legal minority, and therefore unable to vote or be voted for.

From that moment on, the Adult Education Service was created, with the aim of reducing illiteracy in the country, organised by the Adult and Adolescent Education Campaigns from 1947 to 1963. During the campaign, there were prejudiced views about illiteracy, but during the course of the campaign, these views changed, and theoretical pedagogical fields emerged to discuss illiteracy and recognise "that illiterate adults are productive beings, capable of reasoning and solving problems" (NICOLODI, 2011, p.26). Thus, with each passing government, new campaigns and movements emerged to improve adult education and in these transitions of government, the literacy of young people and adults in the 1960s was made up of movements aimed at popular education and culture. These included: Base Education Movement (MEB), Popular Culture Movement (MCP), Popular Culture Centre (CPC) and the Popular Education Campaign of Paraíba (CEPLAR).

Through these educational movements throughout Brazil, debates were promoted in various regions to improve the application and order of adult education. The Law of Guidelines and Bases of National Education (LDBEN), Law 4024/61, was approved, with Article 27 proposing EJA;

> Primary education is compulsory from the age of seven and will only be taught in the national language. For those who start it after that age, special classes or supplementary courses corresponding to their level of development may be formed (BRASIL, 1961, art. 27).

Thus, youth and adult education was gaining a direction for teaching, but it was still criticised for its method of operation involving administrative, educational and financial issues. In the course of these discussions, we highlight one of the main authors of youth and adult education, Paulo Freire. He stands out for bringing a socially committed vision to youth and adult education, aiming for a liberating and problematising education.

In Freire's (1998) view, adult education needs to relate culture and dialogue. His proposal for adult literacy had the basic principle of reading the world before reading writing. Thus, Freire's teaching practice was guided by theoretical and methodological principles related to the development of writing, leading students to assume themselves as subjects of their school and social learning.

In 1967, the EJA took another direction, with the Brazilian Literacy Movement (MOBRAL). This movement arose from the ideas successfully applied in the 1950s and announced the aim of eradicating illiteracy and providing continuing education for adolescents and adults (CUNHA, 1999). However, it was widely criticised for failing to provide and stimulate the student's critical sense.

In 1971, the Lei de Diretrizes e Bases da Educação - LDB - 5692/71, (BRAZIL, 1971) introduced supplementary education, with a specific chapter dedicated to EJA. This law limited the state's duty to the seven to fourteen age group, recognising adult education as a citizen's right. It also extended compulsory basic education from four to eight years, being called primary education, and determined that supplementary education was "to supply regular schooling for adolescents and adults who had not followed or completed it at the proper age" (BRASIL, 1971, art. 24, paragraph a). The education of young people and adults began to be geared towards the exercise of citizenship and better preparation for the labour market. For Ribeiro, Pierro and Joia (2001), this was the first time that EJA was regulated by law and with functions such as: substitution, supply, learning and qualification, thus providing continuing education that could enable the illiterate to become active and critical citizens.

According to Cunha (1999), adult literacy in the 1980s was marked by the spread of research into written language. However, then came the 1998 constitution, which aimed to guarantee compulsory and free primary education for all. The 1988 Constitution brought important advances for the EJA: compulsory and free primary education (BRASIL, 1998). In short, EJA education was offered free of charge to those who did not have access to it at the appropriate age, giving students the right to study and seek autonomy, thus becoming a more critical citizen in society.

In 1985, the Brazilian Literacy Movement (MOBRAL) was replaced by the EDUCAR Foundation, which was replaced in 1990 by the Fund for the Maintenance and Development of Primary Education and Valorisation of the Teaching Profession (FUNDEF).

The 1990s were marked by important movements such as the founding of Forums, the aim of which was to exchange experiences and dialogue on youth and adult education. According to Soares (2004), the Forums are movements that articulate institutions, socialise initiatives and intervene in the development of policies and actions in the area of YAE. It was also important to take part in international congresses such as the World Conference on Education for All, held in Thailand, whose main objective was to show that literacy and post-literacy are interdependent.

In 1996, the Education Guidelines and Bases Law (LDB), Law 9.394/96, came into force, guaranteeing compulsory and free primary and secondary education, including for those who did not have access to it at the proper age (BRASIL, 1996, articles 4 and 37).

In 2000, with CNE/CEB Resolution No. 1 of 5 July 2000, it was established that "The identity of Youth and Adult Education will take into account the situations, student profiles, age groups and will be guided by the principles of equity, difference and proportionality in appropriation and contextualisation" (BRASIL, 2000, art. 5). EJA is thus characterised by the following principles:

> I - with regard to equity, the specific distribution of curricular components in order to provide an equal level of training and re-establish equal rights and opportunities with regard to the right to education;
> II - with regard to difference, the identification and recognition of the specific and inseparable

otherness of young people and adults in their formative process, the valorisation of each
individual's merit and the development of their knowledge and values;
III - in terms of proportionality, the appropriate arrangement and allocation of curricular
components in relation to the specific needs of Youth and Adult Education, with spaces and times
in which pedagogical practices ensure that students have a common educational identity with
other participants in basic schooling. (BRASIL, 2000, art. 5)

Through these principles, teaching in the EJA now shows that everyone has the right to equal opportunities in education, thus recognising the individual merits and values of each student.

As a result, the EJA has been frequently reformulating its curricular proposals. In this sense, important events took place in the 20th and 21st centuries, aimed at improving teaching processes to encourage participation and the resulting qualification of EJA students. In 2005, Resolution 4 of 27 October included a new provision in Resolution 1/2005 of the Brazilian National Education Council (CNE/CEB), which updated the National Curriculum Guidelines defined by the National Education Council for secondary education and technical professional education to the provisions of Decree 5.154/2004. In the same year, CNE/CEB opinion no. 2/2005 introduced the National Youth Inclusion Programme (PROJOVEM), which aimed to provide education, training and community action.

In April 2007, the EJA began to follow a teaching pattern in Brazil, through the Ordinance No. 24 establishes the National Textbook Programme for Youth and Adult Literacy (PNLA). The aim of the programme was to promote compliance with the National Education Plan, which stipulated:

[...] the eradication of illiteracy and the progressive attendance of young people and adults in the
first segment of youth and adult education by 2011 - promoting social inclusion actions,
expanding educational opportunities for young people and adults aged 15 and over who have not
had access to or remained in basic education (BRASIL, 2007, p. 2).

The national programme thus provided textbooks suitable for young people and adults with the aim of eradicating illiteracy throughout the country.

In 2010, the Brazilian Institute of Geography and Statistics (IBGE) carried out the latest and most current survey on EJA, which shows the number of students attending this programme in Santa Catarina. In this survey, attendance at public and private institutions was evidenced, involving the modalities of Secondary Education (1st, 2nd and 3rd year), and Elementary Education (1st to 8th grade).

Table 1: Demographic Census: Education - Secondary Education

People attending secondary school - Private	46.273
People attending secondary school - Public	215.562
People attending secondary school - Total	261.835

Source: 2010 Demographic Census (IBGE)

Table 2: Demographic Census: Education - Primary School

People who attended primary school for young people and adults - Private		4.933
People attending youth and primary education - adults from Public		45.727
People attending youth and primary education - adults from Total		50.660

Source: 2010 Demographic Census (IBGE)

The data presented above shows that there is a significant number of students still attending the EJA. The 2010 census totalled around 312,495 students in private and public institutions.

So, through these data and historical accounts, we realise that the EJA has undergone constant changes with each passing year or decade. However, much can still be researched, but for this to happen it is necessary to establish greater links between school content and students' daily experiences. In addition, teachers and their students need to be valued, so that both can work harmoniously, with the goal of qualifying young people and adults based on school content.

2.2 EJA: TEACHING AND LEARNING

When we study the education of young people and adults, it is necessary to have a partial understanding of the students and their cultural, social and economic diversities. Nicolodi (2011, p. 60), in his research into YAE, demonstrates that:

> EJA students are generally young and adult workers, with professional and life experiences, and for these reasons the EJA teacher cannot proceed in the same way as a regular school teacher; they need to look for teaching methods that take their students' experiences into account and make learning meaningful.

Because they are students with professional experience and are older than students in regular schools, EJA students have "stories to tell, desires and wishes to learn and teach" (SILVA, 2006, p. 57). Through these personal and professional experiences, teachers can relate school content to the daily lives of YAE students, thus giving meaning and significance to the teaching of maths. However, these relationships between the students' stories and school mathematics are often not made due to the traditional teaching model, which focuses on memorising content that has little meaning for the students. This leads us to reflect on how relationships between school knowledge and everyday practices can develop students' creativity and broaden the meaning of school knowledge. In this way, EJA students could have a broader conception of maths and come to understand its importance and use in everyday practices:

> [...] maths teaching needs to provide students with knowledge that favours understanding and transforming their reality. Several efforts need to be made to ensure mathematical learning. The

When students are able to understand that maths makes it easier to solve problems they experience in their daily lives, they may be able to make sense of mathematical knowledge. We know that "the acquisition of mathematical knowledge does not begin for adult learners only when they enter a formal teaching process. This acquisition has been taking place throughout their lives" (DUARTE, 2011, p. 17). For students to be able to broaden their understanding of mathematical knowledge, it is of the utmost importance that teachers use new approaches in their pedagogical practice, as the role of teachers is fundamental in the teaching and learning of maths. According to Melo (2004, p. 37) the teacher:

> [...] it thus assumes a relevant role in the process of constructing the student's mathematical knowledge in the sense that it is up to the student, firstly, to know what, when and how to exploit their prior knowledge; secondly, to decide which prior knowledge should be exploited when approaching new content; and thirdly, to establish relationships between this knowledge (spontaneous or prior knowledge) and school mathematical knowledge (formative knowledge) as a starting point for learning school mathematics.

Maths starts to provide "really meaningful moments for EJA students instead of insisting on hypothetical, artificial and tediously repetitive situations" (FONSECA, 2005, p. 323). This constant search for meaning is important for learning and the formation of critical citizens.

For EJA students to be critically trained, mathematics needs to have "activities that consider exploratory and investigative characteristics, that are systematised, prioritising the procedures developed by the students" (KOORO; LOPES, 2010, p. 4). Mathematics thus needs to be related to everyday activities that stimulate creativity, problematisation and the contextualisation of students' ideas, preparing them for school and personal life.

The Ethnomathematics programme is an alternative "for explaining, learning and understanding, for critically confronting new situations" (D'AMBROSIO, 2004, p. 24). Ethnomathematics seeks to express how mathematical knowledge is produced in everyday life and thus transpose this everyday knowledge into school mathematics. D'Ambrosio (2005) emphasises that the researcher must follow three principles:

o First: Real world to mathematical representations (diagrams, tables, figures).

o Second: Representations with mathematical principles and content.

o Third: Establishing meaning for the student.

Based on these principles, the teacher can, for example, show the students how to use school maths and associate it with their daily lives, such as relating each student's daily spending to purchases made in markets, fairs and commerce in general. In this way, the teacher points out how maths is present in students' daily lives.

Ethnomathematics thus contributes to the political exercise of citizens who are more critical, prepared and have the quality of education to compete in a society that still tries to exclude the less favoured. In the next chapter, in order to give the reader a better understanding of what ethnomathematics is, we will look at the emergence of the programme and its dimensions for teaching.

3

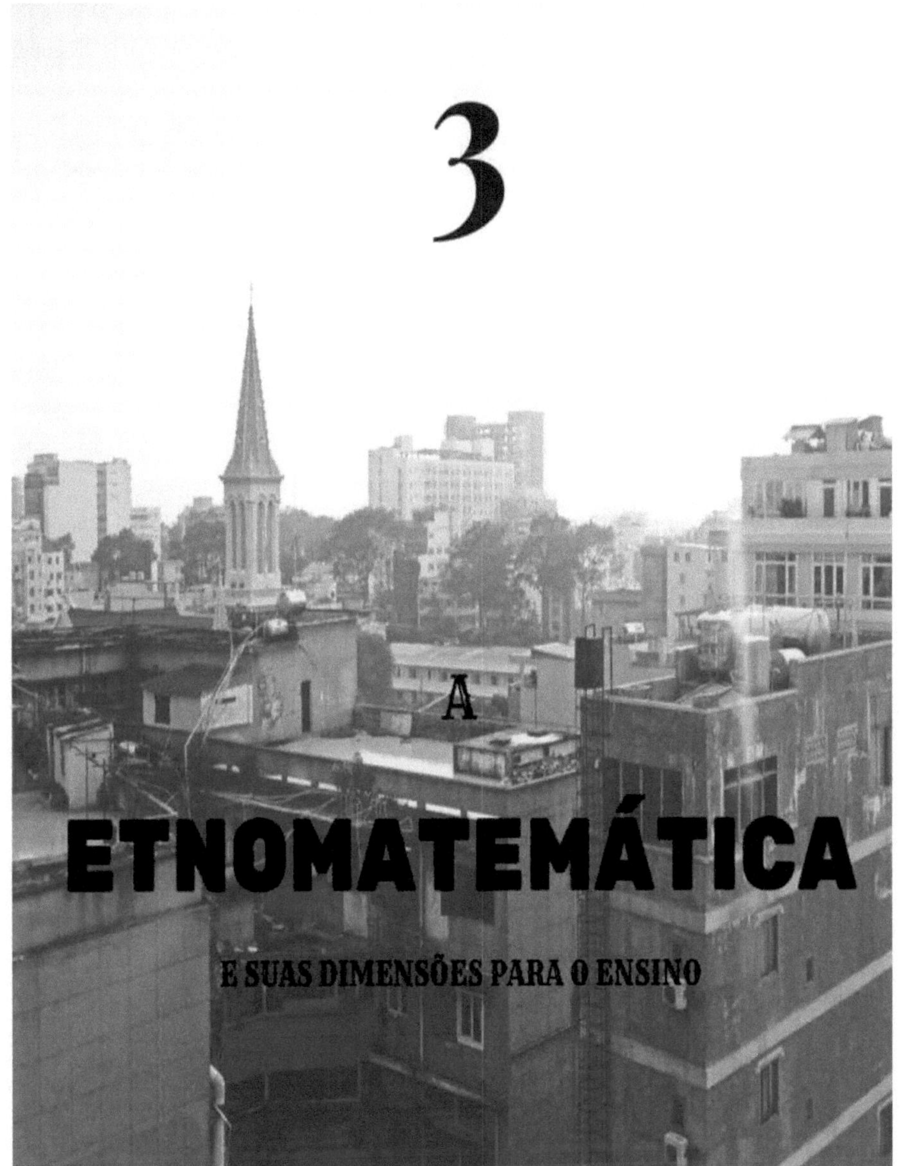

During the history of maths education, the need was felt to expand the traditional paradigms of modern maths, which "aimed to reach only abstract concepts and formulas, such as set theory, algebra and mathematical symbology" (SILVA, SANTOS and BARUFI, 2011, p. 46). During the 1970s, studies were carried out that laid the foundations for a new field of maths teaching: Ethnomathematics. One of the great forerunners of the programme was Ubiratan D'Ambrosio, who was interested in highlighting the knowledge of the social and cultural practices of a particular group (ISOLANI, 2015). From the 5th International Congress on Mathematical Education (ICME-5) held in Adelaide, Australia, in 1984, the term Ethnomathematics was recognised as a research programme with a comprehensive framework.

Ethnomathematics is a field of research with a common nomenclature but different meanings, which is understood according to the conceptions of the researchers who may use it. Miarka & Bicudo (2012) argue that there are different meanings attributed to Ethnomathematics and understand mathematics as a construction based on articulations between different cultures, which is shown in a current that many call "academic mathematics" or "Western mathematics", which is still expanding.

With regard to Ethnomathematics and its relationship with mathematics teaching, D'Ambrosio (1993) suggests that in order to expand the possibilities of Ethnomathematics as a field of research and a proposal for pedagogical action, it is necessary to "free oneself from the Eurocentric pattern and seek to understand, within the individual's own cultural context, their thought processes and their ways of explaining, understanding and performing in their reality" (D'Ambrosio, 1993, p. 5). The great strength of Ethnomathematics lies in understanding the existing cultural dynamics, i.e. reflecting on the cultural aspects of different groups and those that influence them.

D'Ambrósio (1990), Monteiro (2001) and Knijnik (1996) explain how the Ethnomathematics programme came about. For them, Ethnomathematics aims to understand and value mathematics as a cultural production, to relate the knowledge present in the work activities of different social groups, combining it with the solution of everyday problems that are significant to them.

D'Ambrosio offers an example in relation to the start of the Ethnomathematics programme. He reports that:

> When this australopithecine chose and chipped a piece of stone in order to carve a bone, his mathematical mind was revealed. In order to select a stone, it is necessary to assess its dimensions, and in order to chisel it to what is necessary and sufficient to fulfil its intended purpose, it is necessary to assess and compare dimensions. Evaluating and comparing dimensions is one of the most elementary manifestations of mathematical thinking. A first example of ethnomathematics, therefore, is that developed by the Australopithecus **(D'AMBROSIO, 2001, p. 33)**

Since the dawn of humanity, each culture has developed different ways of practising mathematics. Some of these originated in ancient Egypt and ancient Greece. However, other regions of the known and unknown world have also developed significant mathematical ideas and practices. In this way, the Ethnomathematics programme emerged to confront the taboos that mathematics is a universal field of study with no cultural boundaries.

For Knijnik (2004), the discipline of mathematics is in fact an ethnomathematics that originated and developed in Europe in the 16th and 17th centuries, receiving important contributions from the civilisations of the East and Africa for its universalisation and globalisation.

Halmenschlager (2001, p. 15) explains that Ethnomathematics:

> [...] it allows the recognition of different ways of doing mathematics, used by social groups in their daily practices, in an attempt to solve and manage specific realities, which would not always be identifiable from the perspective of academic mathematics

The Ethnomathematics programme aims to understand mathematical knowledge and practice in different cultures throughout human history (D'AMBROSIO, 2002). From this perspective, the study of the Ethnomathematics programme is a "process of generation, organisation and transmission of knowledge in diverse cultural systems" (D'AMBROSIO, 1990, p. 7). From ethnomathematics, it is possible to assign meanings and understand the exchange of experiences in the lives of diverse cultures such as: children, young people, senior citizens, women, the rural world, the seaside, the working class, poor people, and the voices of the third world, among others.

Throughout history, D'Ambrósio (2002) shows that the Ethnomathematics programme was created and developed with a view to reflection, observation and skills techniques. For D'Ambrosio (1990, p. 5 - emphasis added), the Ethnomathematics programme is made up of three terms:

> [...] ethno is now accepted as something very broad, referring to the cultural context, and therefore includes considerations such as language, jargon, codes of behaviour, myths and symbols; matema is a difficult root, which goes in the direction of explaining, knowing, understanding; and tica undoubtedly comes from techne, which is the same root as art and technique

We understand that ethnomathematics will help to develop different techniques for explaining school maths based on students' personal and professional experiences. Based on this assumption, we understand that the teaching of school maths, based on Ethnomathematics, can help students to appropriate scientific knowledge from everyday mathematical knowledge and, in this way, considers the student as the protagonist and not as a spectator of the process. The Ethnomathematics programme takes as its point of reference the epistemologies of each group of people diversified by the method: of cognition and interdisciplinarity, leaving aside concrete thinking and moving on to the abstract, giving rise to what we call culture and historical formation (D' AMBROSIO, 2002), linked to a historiographical proposal that refers to the dynamics of the evolution of actions and knowledge that result from the mutual exposure of cultures.

According to D'Ambrosio (1999, p. 111), "ethnomathematics is a historiographical research programme". For him, the programme covers two specific areas: the internal (intracultural) history: the centrality of science to its development, and the external (intercultural) history: which connects the development of science with the social conditions in which it is produced. The Ethnomathematics programme implies "admitting the importance of the history of science for valuing the historicity of knowledge, where its results constitute historical elements" (BOMBASSARO, 1993, p. 108). However, Ethnomathematics is characterised by mathematical research (level one), with references to ethnology (level two) and maths didactics (level three). Its mathematical movement can be characterised in the following ways: Ethnomathematics and Ethnomathematical Studies. For D'Ambrosio (2002), Ethnomathematics highlights and analyses the influences and social and cultural factors on mathematics teaching, which is treated in a universal way. To describe the importance of the study of Ethnomathematics, Knijnik (2004) presents a framework describing the process of applying Ethnomathematics as a pedagogical resource.

Chart 3 - Ethnomathematics as a Pedagogical Resource

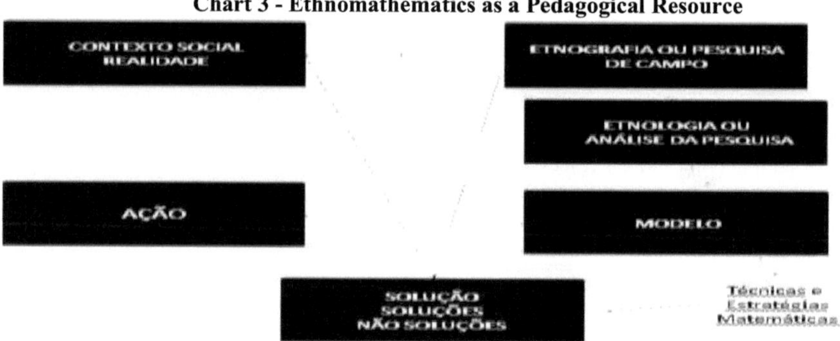

Source: KNIJNIK (2004, p. 80)

In the table above, Knijnik (2004), describes the following steps - action (interaction), social context (everyday life), ethnography (research), ethnology (pros and cons), and finally the aid of techniques that can help create a mathematical model that can be applicable in the student's everyday life.

From Table 3 we can infer that if the school is located in a social context (neighbourhood, region, village), in order to apply ethnomathematics as a theoretical and methodological guideline, the teaching staff needs to have knowledge of the school's social context. In this way, the teaching staff will be able to select mathematical content that is more meaningful to the community. Teachers need to guide their students towards more effective actions within the community's cultural context.

Once the topic has been chosen, we move on to field research, where the data will be collected and then analysed. After the ethnography, the group carries out the ethnology (field analysis). It is at this point that questions and debates arise, seeking a better view of the community and the world. In order to present the work, it is necessary to validate the proposed objectives of the model used and finally to present and universalise the work carried out.

Ubiratan D'Ambrosio (2011), when discussing the universality of Ethnomathematics, not only points to the comparative history of mathematics, associated with cultural anthropology studies, as a kind of antidote against distortions such as arrogance and cultural prejudice, but also states that "the examination of the universality of mathematics is associated with a critical examination of the very institutionalisation of mathematics as a branch of knowledge" (D'Ambrosio, 1998, p. 18). Ethnomathematics thus values culture, the social, in other words, the whole.

According to D'Ambrosio, Ethnomathematics is structured in six dimensions: conceptual, historical, cognitive, epistemological, political and educational. These dimensions are discussed below in order to provide theoretical support for the lessons that were developed during the research.

For Ubiratan D'Ambrosio (2011), the great motivator of the Ethnomathematics programme is seeking to understand mathematical knowledge and practice throughout the history of humanity, shown in different groups, communities or peoples. In this sense, D'Ambrosio (2011, p. 18) argues that:

> [...] every individual develops knowledge and has a behaviour that reflects this knowledge, which in turn changes according to the results of the behaviour. For each individual, their behaviour and their knowledge are in permanent transformation, and they relate to each other in a relationship that we could say is a true symbiosis, in total interdependence

The conceptual dimension of Ethnomathematics shows the practice of human beings in creating, generating and developing new mathematical methods that can demonstrate and amplify events or phenomena observed in their daily lives. For this dimension D'Ambrosio (2011) reports that:

> [...] mathematics, like general knowledge, is a response to the drives for survival and transcendence, which synthesise the existential question of the human species. The species creates theories and practices that resolve the existential question. These theories and practices are the basis for the elaboration of knowledge and **behavioural** decisions, **based on representations of reality (D'AMBROSIO, 2011, p.** 27).

Mathematics can be seen from the conceptual dimension through the acts of explaining, describing and analysing certain events, which human beings need to know, understand and decipher mathematical knowledge because it acts to stimulate their sensory capacity and creativity. The challenges posed by life have led the human species to create theories and practices to resolve issues of existence. These theories become fundamental for representing the reality of each culture, for creating models that respond to the perception of space and time, and for forming knowledge (previous experiences) about reality and the behaviour of individuals.

The historical dimension portrays the past experienced by previous cultures, starting with the Middle Ages, when concrete and quantitative reasoning gave way to qualitative and abstract reasoning, thanks to the emergence of arithmetic and geometry (D'AMBROSIO, 2011). In this sense, this dimension shows that:

<blockquote>
[...] modernity came about with the incorporation of quantitative reasoning, made possible by arithmetic using Indo-Arabic numerals, and later with the extensions of Simon Stevin [decimals] and John Neper [logarithms], culminating in computers [...] More recently, we see an intense search for qualitative reasoning, particularly through artificial intelligence. This trend is in line with the intensification of interest in ethnomathematics, whose qualitative character is strongly predominant **(D'AMBROSIO,** 2011, p. 29).
</blockquote>

Therefore, for this dimension, we are at a moment similar to the Middle Ages. It is believed that the qualitative character tends to be increasingly predominant in the future of man and Ethnomathematics is a manifestation of this new renaissance (D'AMBROSIO, 2011). This makes it necessary to contextualise and problematise mathematics. This dimension also includes the entire development of mathematics and its concepts throughout the evolution of our civilisation. Mathematical knowledge began to be organised in the Mediterranean basin, stemming from the historical interpretation of Egyptian, Babylonian, Jewish, Greek, Roman and other knowledge. In addition, the evidence reveals man's incessant search for a place to live, and to this end he developed intellectual tools to lead him to this end.

In turn, the cognitive dimension is faced with the challenge and need of the human being to establish comparisons and analogies when confronted with new occurrences, interconnecting previous knowledge and relating it to the current knowledge of each student (ISOLANI, 2015). In this sense, this dimension seeks to solve and understand new occurrences, so the student will be attributing and adding new knowledge. According to D'Ambrosio (2011, p. 30)

<blockquote>
[...] mathematical ideas, particularly comparing, classifying, quantifying, measuring, explaining, generalising, inferring and, in some way, evaluating, are ways of thinking that are present throughout the human species. [...] When faced with new situations, we gather experiences from previous situations, adapting them to the new circumstances and thus incorporating new actions and knowledge into our memory
</blockquote>

In this sense, the cognitive dimension of Ethnomathematics seeks to verify the potential for organising knowledge connections between what we have already acquired and what we will experience and acquire, using information that can be shared among other individuals. At the heart of this dimension is man's need to compare, classify, quantify, generalise, infer and even evaluate (D'AMBROSIO, 2011), in other words, it considers and recognises every mathematical manifestation of the human cognitive structure. Ethnomathematics does not devalue the different ways of reasoning and knowledge of other peoples; on the contrary, it validates their strategies for explaining the different events arising from the need for survival and transcendence of the entire human species.

The epistemological dimension works with the transition between the observation of an event and the elaboration of its final resolution. In other words, it works with the relationship between knowing and doing in a culture (D'AMBROSIO, 2011). When dealing with the word epistemology, we are referring to how knowledge is structured, what its phases are and how they are interconnected. In this sense, (D'AMBROSIO, 2011) points out that the individual, being informed, will be able to generate knowledge, being able to express themselves through codes and symbols. Through communication they will be serving their community, assuming the power of knowledge that will serve to explain and deal with reality.

Thus, the epistemological dimension of Ethnomathematics is organised knowledge aimed at explaining the reality that surrounds each student. In order to understand the observation of reality and the theoretical, D'Ambrosio (2001) considers a sequence of three direct questions, which we will transcribe here: 1) How do we move from observations and adhoc practices to experimentation and method? 2) How do we move from experimentation and method to reflection and abstraction? 3. how do we proceed to inventions and theories? These questions guide reflection on the evolution of knowledge and D'Ambrosio (2001) proposes a harmonious cycle of knowledge in an integrated way that considers the constant inter-relationship of the individual with reality and their action.

The political dimension of Ethnomathematics is about valuing the origins of each individual or group of individuals (D'AMBROSIO, 2011). Basically, it seeks to consider the knowledge that each people, race and religion already possess. Ubiratan D'Ambrosio (2011, p. 42) states that:

> [...] each individual carries with them cultural roots that come from their home, from the moment they are born [...] Ethnomathematics fits in with this reflection on decolonisation and the search for real possibilities of access for the subordinate, the marginalised and the excluded.

Thus, the political dimension of Ethnomathematics establishes the preservation of the knowledge acquired by the individual from birth, throughout their life, through communication and interaction with other human beings. In our colonisation, the Jesuits performed this function with great skill: they imposed their language, customs and religion, and worked efficiently to inferiorise the native people. However, Ethnomathematics recognises and values the thinking of other cultures - it doesn't remove each individual's reference point, but reinforces their own roots; it doesn't end with a selective practice, but restores the dignity of each individual and works on the process of transition from subordination to autonomy.

And finally, the educational dimension relates to the use of informal everyday maths to contribute to the teaching of academic maths. The educational dimension for D'Ambrosio (2011, p. 42-43): "is not about ignoring or rejecting modern knowledge and behaviour. But rather to improve them by incorporating values of humanity, synthesised in an ethic of respect, solidarity and cooperation". In other words, it's about relating everyday maths to the maths taught in schools. Ethnomathematics brings a pedagogical proposal for doing maths that deals with real situations of time and space and that considers the importance of different cultures and traditions for the formation of a new society: transcultural and transdisciplinary.

In this way, the educational dimension means that academic content must be related to students' previous content in order to establish greater meaning in teaching and learning.

The ethical dimension of Ethnomathematics is also important in research into Youth and Adult Education. Questions such as "What knowledge is exposed in this culture?" and "Why do we study it?". A first answer to these questions could be that ethnomathematics develops an ability to resist the domination of cultures (MIARKA, 2011). In this sense, this dimension enables a different way of thinking and acting about maths in everyday life.

For Miarka (2011, p. 31), the ethical dimension of Ethnomathematics serves to point out the "relevance of work that seeks to understand the various currents of Ethnomathematics, indicating their approximations and divergences".

Through these dimensions, we realise that Ethnomathematics is not a teaching method, but a proposal that stimulates the development of creativity, leading to new forms of intercultural relations. This is confirmed in this quote: "it is a programme that aims to explain the processes of generation, organisation and transmission of knowledge in various cultural systems and the interactive forces that act in and between the three processes" (D'AMBROSIO, 2001, p. 24).

We therefore understand that maths experienced in the cultural and social context of each student can contribute to broadening their understanding of reality and the world. But for this to happen, it is essential to interact school knowledge with the students' everyday practices. If this is not possible, then maths is just a way of solving questions without meaning or significance. From this perspective, we believe that one of the ways to support this is through pedagogical actions in maths teaching that are geared towards the socio-cultural context of each EJA student.

CHAPTER 4

APPLIED ACTIVITIES: EXAMPLES

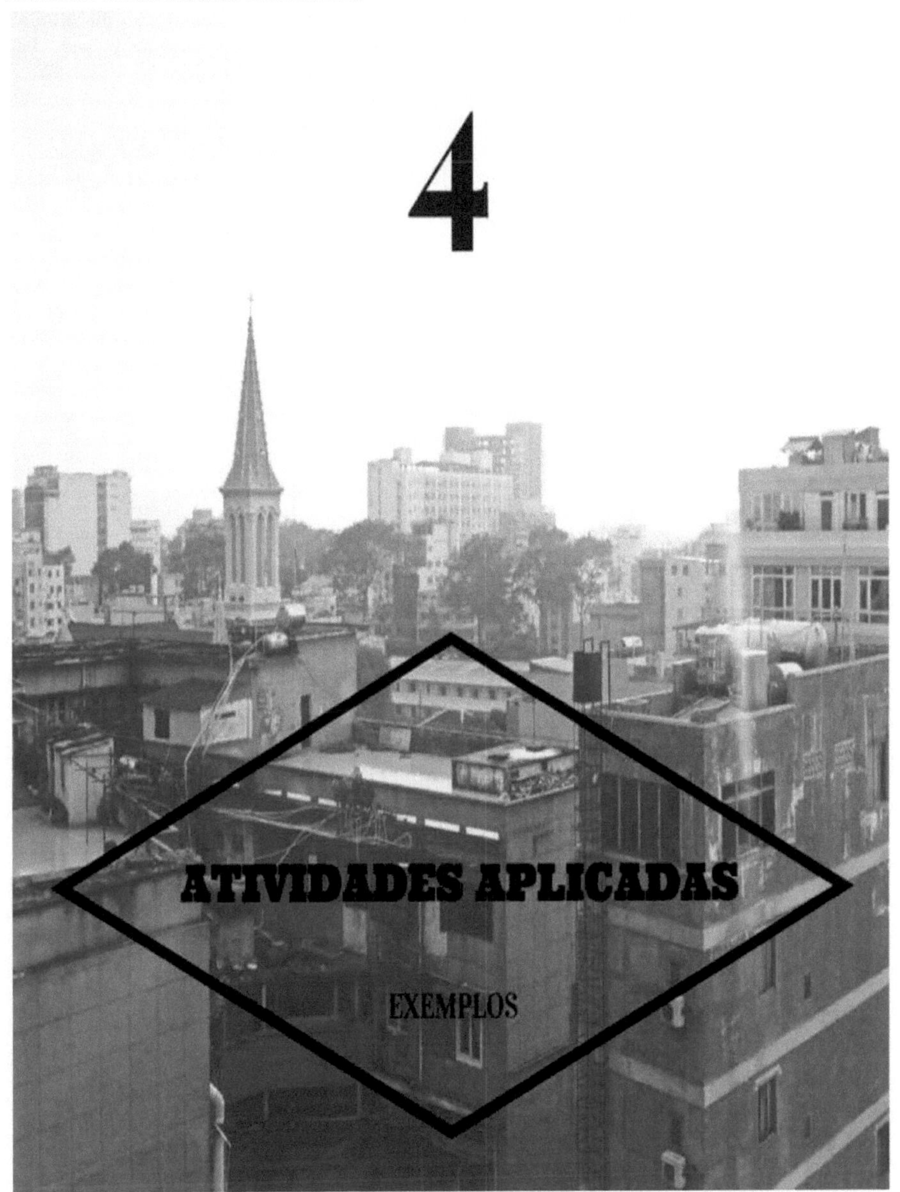

This is a qualitative, participant-based study. The teaching unit surveyed was the Núcleo Avançado de Ensino Supletivo (NAES) in Itapema - SC, with twenty students in the third year of secondary school.

The municipality of Itapema is located at the mouth of the Itajaí-Açu River (Figure 2) and belongs to the Itajaí micro-region in the state of Santa Catarina, Brazil. Founded on 21 April 1962, its geographical area is 59.022 km^2 , with an estimated population of 45,814 (IBGE/2010 Census), known as the Capital of Ultralights (ITAPEMA, 2013).

NAES, part of the Santa Catarina public school system, was created in 1998 and caters for primary and secondary school students. It is located in the Meia Praia neighbourhood, at 428, 230th Street. In the year of the research, 2014, it had 300 students and the minimum age to attend this type of education is 15 and the duration varies from one and a half to three years.

The subjects surveyed are aged between 18 and 50. There are twenty-one students in the third year of secondary school, three of whom work in construction, one is a carpenter with five years' experience in construction, one is a master builder with four years' experience, one is a servant with two years' experience and others who had no experience took part in the process of explaining, building and accepting the activities and acceptance.

For the research we used data collection instruments such as semi-structured interviews, which were recorded on video and later transcribed and analysed. The interviews were carried out before the research with students who had experience in the construction industry and then students without experience in the construction industry were also interviewed, after which the construction activities began.

All the activities carried out in class were aimed at investigating/demonstrating how maths is used in the construction industry, in an attempt to relate everyday maths to school maths in order to teach the mathematical concepts set out in the content plan for the EJA class, the subjects of this research. The proposed activities followed the teaching plan for the third year of the EJA secondary school

The chronological continuation of the activities carried out in class is shown below. The activities were developed according to the context and interest of the students with experience in construction, gathered from the initial interview with all the students in the class. During the meetings, the lessons were organised in such a way that each group worked on solving the activities requested by the teacher, starting from the topic of interest to the students and taking into account everyone's previous knowledge.

The activity on building a water tank (Activity 1) aimed to show the process of using maths to build a tank in construction and its related measurements.

The activity of drawing up a budget (Activity 2) investigates the process of planning the purchase of materials needed to build a wall and a pavement, and for all budgets it is necessary to research companies specialising in construction.

To finalise the activities, we built a staircase (Activity 3), in which the student was faced with a problem situation in which the buyer of the property asked the master builder about the measurements of a staircase and what type of standard staircase could be used in the construction industry. In the activities to build the water tank (Activity 1) and the staircase (Activity 3), full-scale mock-ups were made in order to utilise the knowledge of students with experience in the construction industry in the classroom.

An open-ended questionnaire can also be considered a data collection tool. In this research, the questionnaire was applied at the end of the lessons as a way of checking the level of learning and/or satisfaction of the class after the curricular unit worked on.

Finally, we used video and audio recordings of the activities carried out in the classroom. The video and audio recordings were transcribed in the form of a summary of each meeting with the students, explaining the content worked on, the methodology used by the teacher-researcher and the debates between the students and the teacher. The following is a preliminary account of the data collected, related to the central question of the research.

4.2 MATHEMATICAL CONCEPTS USED IN CIVIL CONSTRUCTION BY EYA STUDENTS AND THE RELATIONSHIP WITH ETHNOMATEMATICS

Initially, in order to draw up the units of analysis for the research, we tried to detect the prior knowledge of three students in the class who work in the construction industry. Based on the reports of the students with experience in the construction industry, indicated here with the letters of the alphabet (F, G and K), and other students who do not have experience in the construction industry will be indicated by the letters of the alphabet (A, B, C, D,...) to preserve their identities, it was possible to identify some dimensions of Ethnomathematics present in the process (Chart 5).

Chart 5 - Students' Conceptions of Maths

■ UNIT OF ANALYSIS 1 - MATHEMATICAL CONCEPTS USED IN CONSTRUCTION	DIMENSIONS OF ETHNOMATHEMATICS IDENTIFIED
K: "I use it to give the final value to customers, my father taught me like that, a pavement, and side by side times my service. And the material depends on the size, right?	Historical
F: "These measurements *are learnt* in practice, on a daily basis, what 90° degrees and 45° degrees are and so on. Also dividing the parts, the radii and the diameter, and calculating the area for each one. For me, maths is important in my day-to-day life, without these measurements I can't do anything."	Politics

G: **"Firstly, I chose a measure *that* was proportional to the real size that I use on a daily basis,** the measure came from the square, which is two lines and that closed the square at an angle of 90° degrees and here we pulled 13 cm from the square and 21 from the square to here and then the centre of the diameter and there were two radii of 17.2 cm and with this diameter we **made the semicircle** ...**In this case it's 17.2 cm times 17.2 cm, which** that's 295.84 times 3.14 which is 9 metres and 28 cm is the area we have here inside **our circumference and the total here is part of a circle that has 180 degrees".**

Source: Author's personal archive

These reports collected in the interviews reveal that the students' daily lives are full of mathematical information that can help them in their day-to-day work.

In the comments above, the student divides any figure into proportional parts, and he knows that every figure starts from a 90° angle, so subdivisions are necessary to find measurements and angles. It is also clear that the student already has some experience in the construction industry and that the mathematical content studied and related in the classroom makes sense to him. In this way, the students end up showing their professional experiences and exposing their relationships with school mathematics, contributing to a more dialogued lesson that involves everyday practice and school mathematics, thus establishing meanings and significance about the mathematical content taught. Thus, we can say that daily interaction with practice is the starting point for educators to be able to find meaning between everyday maths and school maths.

Another important report came from student G. when he was asked what types of examples or calculations are used in the construction of a water tank, building estimates, building a staircase, in his professional practice. Analysing this interview, it was possible to see that the student uses mathematical concepts in his professional practice, such as: right and shallow angles, radius, diameter and circular area ($A = n.r^2$). For the student, the semicircles are made in an aluminium sheet with proportional cuts, which will be the base of the steps of the spiral staircase. His calculations are then carried out in a similar way to school maths, as the student measures the diameter, then multiplies the square radii by pi, thus making it possible to know the amount of material that could be used for a spiral or traditional staircase.

During the interview with student K, who has professional experience in the construction industry, the teacher asked him how he applies maths concepts in his practical experience.

Student K then said: "I use it, to give the final value to clients, my father taught me like this, a pavement, and side by side times my service. And the material depends on the size, right? This situation shows that the student learnt from his father how to calculate and build a budget and continues to apply this method today. As a result, student K relates the maths that has been passed down from generation to generation, using school maths resources such as: the area of square figures (Plane Geometry) and operations between factors and products, thus enabling a final value for the budget. This implies the historical dimension of Ethnomathematics. For D'Ambrosio (2011), when the student portrays knowledge from the past, which is passed down from generation to generation.

The interviews mentioned and interpreted above guided the choice of content in the maths lesson plans for the aforementioned EJA class, the object of study in this research, such as: areas of quadrangular and circular figures, angles, trigonometry and volumes. The lessons and their activities/results corresponding to the EJA teaching plan, under an Ethnomathematics approach, are presented and discussed in item 4.2 below.

4.3 MATHS ACTIVITIES DEVELOPED AND THE PRINCIPLES OF ETHNOMATHEMATICS PRESENT IN PEDAGOGICAL PRACTICE

The curricular unit worked on had the following objectives:

- Use mathematical concepts about measurements, angles and volumes used in construction to learn school maths concepts.
- Identify the principles of Ethnomathematics present in the pedagogical practice applied to the teaching of school maths concepts.

The first activity aimed to identify the concepts of geometry, angles and volume measurements present in the construction of a water tank. The students used materials such as pencils, rulers, cardboard boxes, plastic bags, calculators and waterproof sheets to build the water tank. The total time spent on the activity was 9 hours, with lessons lasting 60 minutes each.

The purpose of activity 2 was to ask the students to make a connection between the mathematical concepts they use when buying building materials and school maths. In this way, it was possible to identify whether the students already had experience of making budgets and the mathematical concepts involved. To carry out this activity, the students first had to identify the types of building materials needed to make a pavement and a wall, and the purchase prices for each item. After discussions and explanations, the teacher asked the students to draw up a budget and handed out the activity to each of them. The following materials were needed for this activity: calculator, paper, cardboard box, leveller, tape measure and pencil. The total time spent on the activity was 15 hours, with lessons lasting 60 minutes each.

Finally, activity 3 aimed to identify the procedure adopted by the students to build a staircase, and also to identify the trigonometry concepts present in the construction of a staircase in a building. The following materials were needed for this activity: pencil, calculator, ruler, plumb line, compass, glue and cardboard boxes of different sizes. The activity took 15 hours to complete, with lessons lasting 60 minutes each.

In all the activities proposed, we were concerned to understand the possible difficulties encountered by the students and asked them to build small-scale models, especially in activities 1 (one) and 3 (three). However, all the activities and meetings (Table 6) were related to school mathematics and linked to the students' everyday lives, highlighting the importance of contextualising and problematising the content in an attempt to develop maths teaching that aroused interest and made sense to the students.

Table 6 - Example for Application

<table>
<tr><td colspan="2">Youth and Adult Education-EJA NAES- Itapema</td></tr>
<tr><td colspan="2">Evening - 6.30pm to 9.30pm</td></tr>
<tr><td>Subject:</td><td>MathematicsHours: 42h (14 meetings)</td></tr>
<tr><td colspan="2">Meeting 1- Submission of the informed consent form to the unit coordinator. Presentation of the dissertation project to the students, explaining the whole process of applying the research, the activities and evaluations to be carried out.</td></tr>
<tr><td colspan="2">Meeting 2 - Collection of the signed informed consent forms regarding the students' participation in the dissertation research.</td></tr>
<tr><td colspan="2">Meeting 3 - Activity 1: Explanation of the manufacture of a water tank by students with experience in construction, discussions, explanation of school maths and resolution of examples of volumes by the teacher. A standard water tank will be established and during the explanations there will be dialogue between the construction workers and the other students.</td></tr>
<tr><td colspan="2">Meetings 4 and 5 - Making scale models of the water tank, student presentation and assessment.</td></tr>
<tr><td colspan="2">Meeting 6 - Activity 2: Students with experience in the field carry out a construction budget. Indication of the materials needed to build pavements and walls. During the explanations, there will be dialogue between the construction workers and the other students.</td></tr>
<tr><td colspan="2">Meeting 7 - Explanation by the teacher of: the concepts of square areas, the cost of items such as bricks, stone, sand, iron, cement, labour and how to carry out the calculation procedures up to the final budget.</td></tr>
<tr><td colspan="2">Meetings 8 and 9 - Preparation of the budgets, presentation by the students and evaluation through debates and presentation of ideas.</td></tr>
<tr><td colspan="2">Meeting 10 - Activity 3: Explanation by construction workers about stair construction, stair pattern, angles. During the explanations there will be dialogue between the construction workers and the other students.</td></tr>
<tr><td colspan="2">Meeting 11 - Explanation by the teacher of the concepts of trigonometry, angles (acute obtuse and internal), and activities related to the content of school maths.</td></tr>
<tr><td colspan="2">Meeting 12 - Revision of the concept of trigonometry, to find the third measure of the ladder and activities related to the content.</td></tr>
<tr><td colspan="2">Meetings 13 and 14 - Making scale models of the staircase, presenting them to the students and assessing them.</td></tr>
</table>

4.3 ACTIVITIES TO DEVELOP IN THE CLASSROOM

4.3.1 Interview

<u>Description</u>: Interviews with students with or without experience in the construction industry. The teacher will ask questions following a pre-established script, the theme of which will be their daily practice, and will observe how the answers are provided. The teacher will act as a mediator of knowledge, leading the dialogue.

Interview on Mathematical Concepts

1- HOW DO YOU APPLY MATHS CONCEPTS IN YOUR PRACTICAL (PROFESSIONAL) EXPERIENCE?

2- EXPLAIN, WITH EXAMPLES, HOW YOU CALCULATE THE **CONSTRUCTION OF A WATER TANK, BUDGETS FOR BUILDING WORK, THE** CONSTRUCTION OF A STAIRCASE, IN YOUR PROFESSIONAL PRACTICE?

<u>Materials needed</u>: a script of questions and the use of a video camera and a photographic camera to collect the students' testimonies.

<u>Objectives</u>:

a) Conduct a survey of the students' knowledge of their everyday practices, even before the teacher starts working on the subject content;

b) Encourage the collective participation of students to collect testimonies.

<u>Estimated time</u>: 2 (two) meetings of 60 (sixty) minutes each.

4.3.2 Building a water tank

<u>Description</u>: To begin the activity, it is necessary to have an explanation with students who have experience in the construction industry, so that through discussion the teacher and the students can ascertain what types of water tanks are currently used in the construction industry. After this stage, the teacher should work with the students to establish a water tank standard. In this way, the teacher will be able to apply relationships between the mathematical concepts applicable to the construction of a water tank and the geometry and volume measurement content to be learnt in school maths. Next, the teacher will give each student the activity and ask them to assemble a miniature water tank on a reduced scale.

<u>Objectives</u>: To identify the concepts of geometry and volume measurements present in the construction of a water tank: To identify the applicability of school mathematics in everyday construction work and to recognise the importance of the concept of geometry and measurements in

everyday life.

<u>Contents:</u> Volume.

<u>Estimated time:</u> 3 (three) meetings of 60 (sixty) minutes each.

<u>Materials needed:</u> pencil, ruler, cardboard boxes, plastic bags and waterproof sheets.

Water tank construction

Name:____

During an inspection related to the purchase of a house, the buyer, together with the master builder, realised that the square water tank was leaking. After a brief conversation, they decided to renovate it.

The box will have the following measurements: two metres long, one metre wide and two metres high.

However, the buyer had doubts and asked: How many litres will fit in the water tank? Based on your knowledge of maths, demonstrate the answer to the buyer in the spaces below, indicating the calculations used.

4.3.3 Realising a Budget

<u>Description:</u>- For this activity, it is necessary for construction students to give a previous account of their experience in buying materials for a budget involving construction, after which the students will be asked about the values of each product, labour and the possible quantity of materials that can be used. In this way, the teacher, together with the students, can make a connection between the mathematical concepts applicable to the purchase of construction materials and school maths.

<u>Objectives:</u> To identify the applicability of school mathematics in everyday life in the construction industry; To recognise the importance of the concept of geometry and measurements in everyday life and to identify the operations required to purchase products such as bricks, sand, cement, tiles, flooring and levelling buildings.

<u>Contents</u>: Area, measures and daily operations.

<u>Estimated time</u>: 4 (four) meetings of 60 (sixty) minutes each.

<u>Materials needed</u>: pencil, ruler, cardboard box, quiver, leveller and calculator.

Budget realisation

Name:____

You have been hired to carry out an estimate. The owner would like to refurbish the pavement (photo 1) of the house and build a new wall (photo 2). The driveway is two metres long by four metres wide, and the wall is two metres high by five metres wide. The owner has set a spending limit of around R$2,000.00 (two thousand reais).

So how much could you charge to do this job?

Justify this by showing the materials needed and their quantities, as well as the calculations in the budget for the pavement and wall?

Figure 1 - Pavement model Figure 2 - Wall model

Fonte: www.forceclean.com.br   Fonte: www.decorandocasas.com.br

____
____
____
____
____
____
____
____
____
____
____
____
____
____
____

4.3.4 Building a staircase

Description: To carry out the activity, the teacher will first ask the students to explain how to make a ladder in their everyday lives. After the explanations and discussions, the teacher can ask the students to relate the maths concepts used to build the ladder to the trigonometry and angles content to be learnt in school maths. In this way, the teacher can propose the construction of a small-scale model in order to bring a little more of the student's reality into the classroom.

Objectives: To identify the applicability of school maths in everyday construction work; To investigate and identify the trigonometry concepts present in the construction of a building staircase.

Content: Trigonometry and angles applicable to construction.

Estimated time: 5 meetings of 60 minutes each.

Materials needed: Pencil, ruler and protractor, trigonometric table, lines, feather, compass and cardboard box.

Building a staircase

Name: ____

A seller offered a two-storey apartment to a buyer, who was interested in the price and payment terms. However, when he went to visit, he came across a staircase under construction. Analysing the situation, the buyer asked the construction manager what the measurements of the stairs would be. He told him that it would be built to the following dimensions: 18 cm long and 27 cm high. However, he was having trouble finding the final measurement (third measurement).

Based on the knowledge of trigonometry learnt in maths lessons, what is this third measure and which angles are used?

Figure 1 - Ladder construction

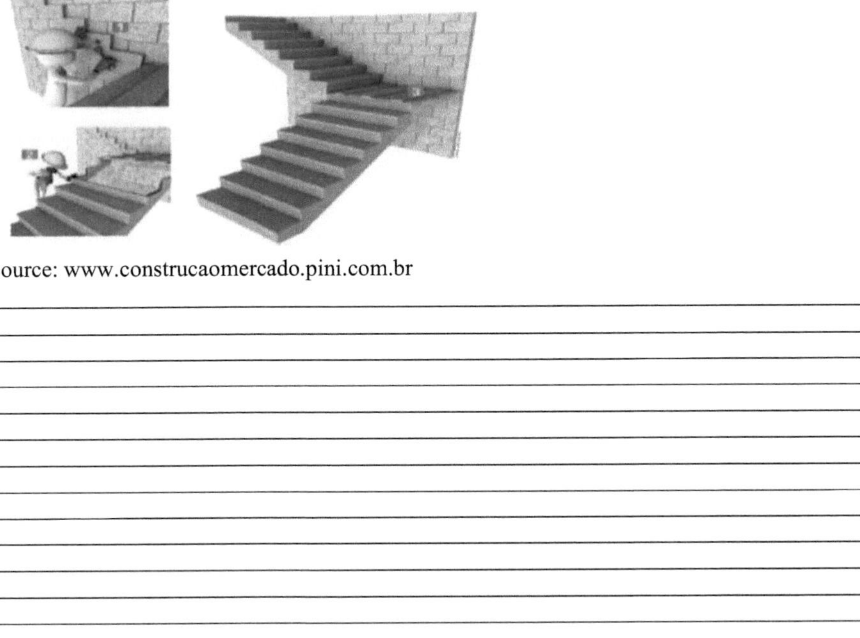

Source: www.construcaomercado.pini.com.br

CHAPTER 5

FINAL CONSIDERATIONS

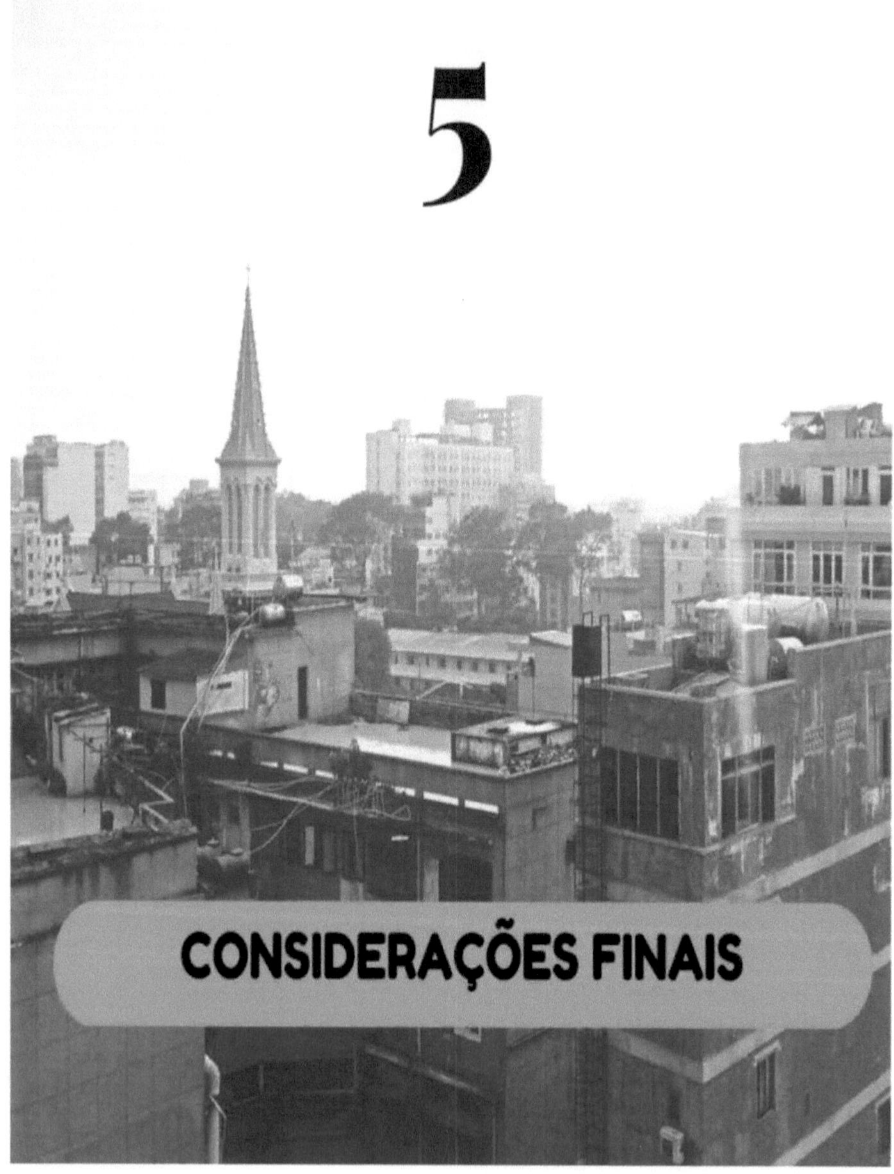

The general aim of this research was to identify the contributions of ethnomathematics to a maths teaching process for young people and adults, which linked everyday building practices and the teaching of school maths.

In order to answer this question, we chose to address content that was of interest to the students and that could be used to explore mathematical content in the EJA. After an informal discussion with the students, we chose the following content: building a water tank, making a budget and making a ladder. We believe that, in general, all the objectives set were successfully achieved, given that the analysis and presentation of the academic content made more sense to the EJA students.

In this sense, we can say that the activities carried out, guided by the dimensions of Ethnomathematics, can be used as innovative and creative methodological resources for the pedagogical practice of applied school mathematics in the EJA, as they promoted more effective participation in maths classes.

All of the activities developed in the classes, guided by the theoretical and methodological assumptions of Ethnomathematics, served as methodological strategies to make school maths classes more meaningful and which aroused the creativity and interest of the EJA students. Based on the activities carried out in the classroom, we analysed the students' difficulties and limitations. The students' experiences showed that school maths can be associated on a daily basis with their professional or everyday practices. The activities carried out during the research process showed results that were achieved thanks to the collective and collaborative participation of the students and teacher. Based on the debates presented in the research, we can outline an innovative strategy for teaching school maths. In this sense, Ethnomathematics has shown that the practice of dialogue based on the students' knowledge and the topics to be studied has contributed to the process of teaching and learning school mathematics.

From this research, it was possible to awaken the interest of students in Youth and Adult Education in the study of school maths. We realised that it is possible to make students understand that mathematical knowledge is found far beyond the classroom.

Ethnomathematics has shown that the daily experiences of a group of professionals can be associated with the study of school maths, as long as they are worked on within an educational context of basic, vocational and higher education, relating content such as: volume, units of measurement used in everyday life, trigonometry, angles, areas, among others.

It was clear from the research that the students had some experience in the construction industry and that the mathematical content studied and related in the classroom made sense to them. We can therefore say that daily interaction with practice is the starting point for educators to be able to find meanings and relationships between the social and cultural experiences of each student and school maths. Furthermore, in the teaching process that establishes a relationship between school maths and everyday maths, it is clear that EJA students implicitly carry out calculations related to Plane and Spatial Geometry. In this case, we can say that mathematical calculations are essential for the construction industry.

The research revealed a gap between the students' prior knowledge related to the everyday practices of construction workers and school mathematical knowledge. Based on this, we tried to bring these perceptions closer together with a pedagogical practice based on the principles of Ethnomathematics, using as a reference the following dimensions: conceptual, historical, cognitive, epistemological, political and educational.

Therefore, this work contributes to the teaching of natural sciences and mathematics, enabling the growth of responsibility, collaboration, persistence, organisational skills, creativity and respect for different cultures, allowing construction workers to be the protagonists of learning.

6

BORBA, O. F. Aspectos teóricos da pesquisa participante: considerações sobre o significado do papel da ciência na participação popular. In: BRANDÃO, C. R. (Org.). Pesquisa Participante. 7 ed. São Paulo: Brasiliense. 1988. p. 20-25.

SECRETARIAT OF FUNDAMENTAL EDUCATION. BRAZIL. National Curriculum Parameters. Brasília, 1997.

SECRETARIAT OF FUNDAMENTAL EDUCATION. BRAZIL. National Curriculum Parameters. Brasilia, 1998.

BRAZIL. Ministry of Education. Department of Basic Education. Curriculum Proposal for Youth and Adult Education: second segment of primary education (5^a to 8ª grade). Brasília: Secretariat of Basic Education, 2002. 240 p. Vol. 3 - Maths.

BRAZIL. The National Textbook Programme for Youth and Adult Literacy (PNLA). 2007. Available at: <http://portal.mec.gov.br/index.php?option=com content&view=article&id=12386:pnla&cati d=314:pnla&Itemid=636>. Accessed on: 18/05/2015.

CORRÊA, C. M. de Sena. Fishing net: a mediating element for teaching geometry. 2000. 195f. Master's dissertation. Florianópolis - Santa Catarina. 2000.

CUNHA, Conceição Maria Da. Introduction - discussing basic concepts. In: SEED-MEC. Leap into the future: youth and adult education. Brasília. 1999. p. 40-50.

D'AMBRÓSIO, Ubiratan. Ethnomathematics: the link between traditions and modernity. Belo Horizonte: Autêntica, 2001.

D'AMBROSIO, Ubiratan. Ethnomathematics. 2. ed. São Paulo: Àtica, 1990.

D'AMBRÓSIO, U. The Conceptual Bases of the Ethnomathematics Programme. Latin American Journal of Ethnomathematics, Brasília, Vol.7.n.2. p.100-107, 2014.

D'AMBRÓSIO, U. Ethnomathematics: a programme. A Educação em Revista-SBEM, Blumenau. v.1. n.1, p.5-11. 1993.

D'AMBRÓSIO, U. Educação matemática: da teoria à prática. 2 ed. Campinas: Papirus. 1998.

DUARTE. C. G. Ethnomathematics and social practices in the construction industry. Unisinos - Rio Grande Do Sul. 2011.

FONSECA, M. C. F. R. Maths education for young people and adults: specificities, challenges and contributions. Belo Horizonte: Autêntica. 2005.

FRANCO, I. C. A. Multiplicative procedures: from mental calculation to school representation

FREIRE, Paulo. Pedagogy of the Oppressed. 43. ed. Rio de Janeiro, Editora Paz e Terra, 2006.

GODINO, J. D. & BATANERO, C. (1998). Clarifying the meaning of mathematical objects as a priority area of research in mathematics education. In, A. SIERPINSKA, A. & KILPATRICK, J. (eds.), Mathematics Education as a Research Domain: A Search for Identity (pp. 177-195). Dordrecht: Kluwer, A. P, 1998.

GLAZIER, Jack D. & POWELL, Ronald. R. Qualitative research in information management. Englewood, CO: Libraries Unlimited, 1992.

HALMENSCHLAGER, Vera Lúcia da Silva. Ethnomathematics: An educational experience. São Paulo: Summus, 2001.

ISOLANI. L. G. Principles of ethnomathematics applied to the process of teaching electricity in a vocational course. 2014. 87f. Master's dissertation. Blumenau. Santa Catarina, 2014.

ITAPEMA. Year 2012. Available at: <www.itapema.sc.gov.br>. Accessed on: 23 September 2014.

KAPLAN, Bonnie; DUCHON, Dennis. Combining qualitative and quantitative methods in information systems research: a case study. MIS Quarterly, v. 12, n. 4, p. 571-586, Dec, 1988.

KNIJNIK, Gelsa. Ethnomathematics, curriculum and teacher training. Santa Cruz do Sul: Ed. UNISC, 2004.

KILPATRICK, J. Pulling up stakes: an attempt to demarcate maths education as a professional and scientific field. Campinas, SP: Zetetiké, vol. 4. n° 5, 1996.

MELO, Maria José M. D. Do "contar de cabeça" à cabeça para contar: historias de vida, representações e saberes matáticos na educação de jovens e adultos. Dissertation (Master's in Education). Natal: UFRN. 2004.

MIARKA, Roger; BICUDO, Maria Aparecida Viggiani. Mathematics in/on/or Ethnomathematics? Article. Revista latino americana de etnomatemática, Vol.5. n°1, p.149-158, 2012.

MIARKA, Roger. Ethnomathematics: from the ontic to the ontological. Thesis (doctorate) - Universidade Estadual Paulista, Instituto de Geociências e Ciências Exatas Rio Claro, 2011.

MONTEIRO, Alexandrina; POMPEU JR, Geraldo. Maths and cross-cutting themes. São Paulo: Moderna, 2001.

NICOLODI, R. The teaching of maths in youth and adult education: an approach based on didactic sequences. 2011. Master's dissertation. Blumenau - Santa Catarina, 2011.

PORCARO, Rosa Cristina. The history of youth and adult education in Brazil. Year 2006. Viçosa: Department of Education, Federal University of Viçosa, 2007.

RIBEIRO, V. M. Educação de jovens e adultos: proposta curricular para o 1° segmento do ensino fundamental. São Paulo: Ação Educativa; Brasília: MEC, 1997.

SANTOS, R. F. dos. Participatory research: what it is and how it is done. Year 2012. Available at <http://baixadacarioca.wordpress.com/2012/03/19/pesquisa-participante-o-que-e-como-se- faz/>. Accessed on: 13 October 2014.

SILVA, M. A. D. Ethnomathematics in an EYA classroom: the bricklayer's experience, 2007. 213f. Dissertation (Master's) - São Paulo, 2007.

SOARES, Leôncio José Gomes. The emergence of YAE forums in Brazil: articulating, socialising and intervening. In: RAAAB, alfabetização e Cidadania - políticas Públicas e EJA. Revista de eja, n.17, 2004.

SMOLE S. K; DINIZ, I. M; MILANI.E. Jogos de matemática de 6 ao 9 ano. Artmed. Porto Alegre,

2007.

yes I want morebooks!

Buy your books fast and straightforward online - at one of world's fastest growing online book stores! Environmentally sound due to Print-on-Demand technologies.

Buy your books online at
www.morebooks.shop

Kaufen Sie Ihre Bücher schnell und unkompliziert online – auf einer der am schnellsten wachsenden Buchhandelsplattformen weltweit! Dank Print-On-Demand umwelt- und ressourcenschonend produzi ert.

Bücher schneller online kaufen
www.morebooks.shop

info@omniscriptum.com
www.omniscriptum.com

Printed by Books on Demand GmbH, Norderstedt / Germany